MW01469923

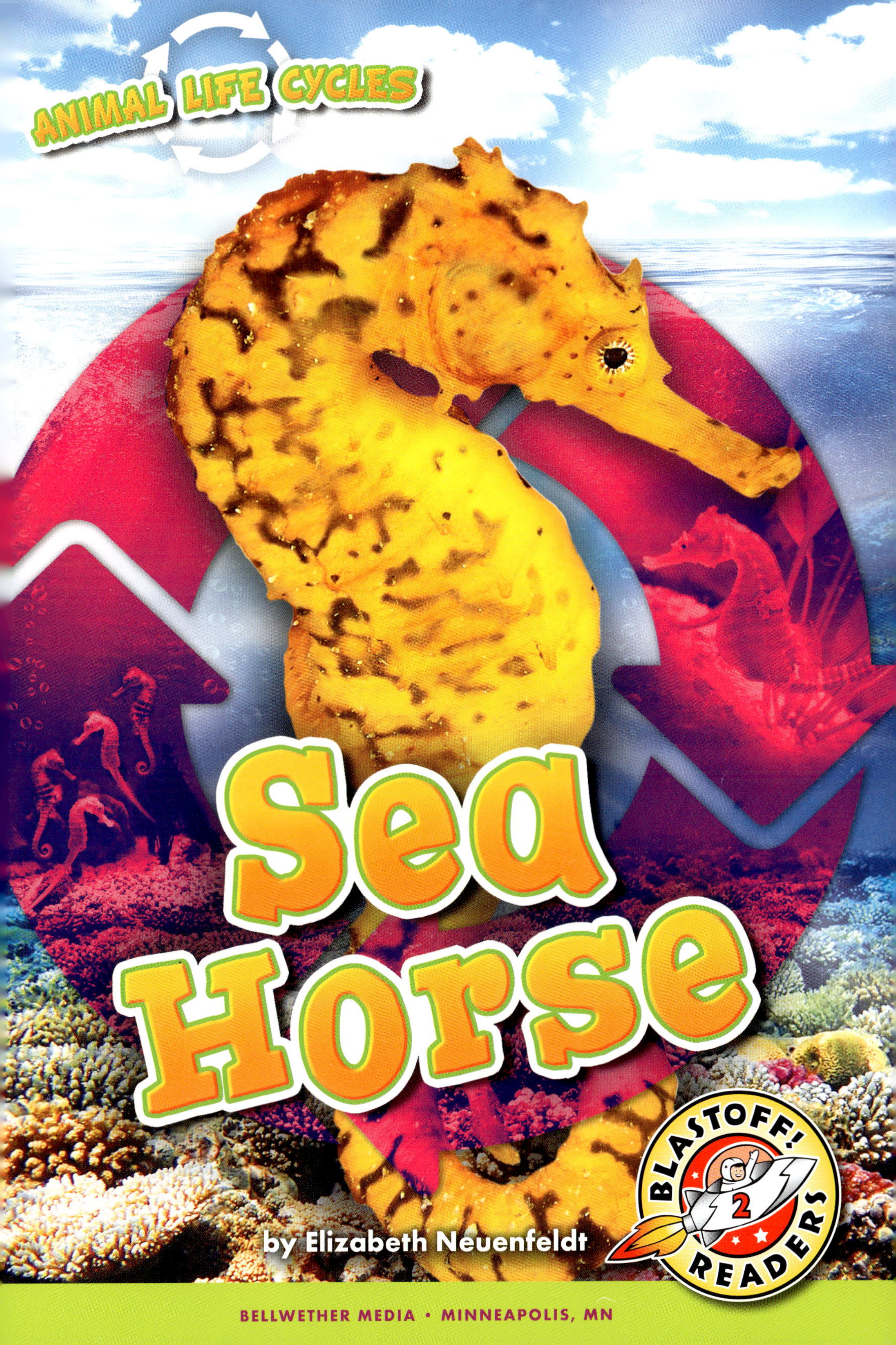
ANIMAL LIFE CYCLES
Sea Horse
by Elizabeth Neuenfeldt
BLASTOFF! READERS
2
BELLWETHER MEDIA • MINNEAPOLIS, MN

Blastoff! Readers are carefully developed by literacy experts to build reading stamina and move students toward fluency by combining standards-based content with developmentally appropriate text.

Level 1 provides the most support through repetition of high-frequency words, light text, predictable sentence patterns, and strong visual support.

Level 2 offers early readers a bit more challenge through varied sentences, increased text load, and text-supportive special features.

Level 3 advances early-fluent readers toward fluency through increased text load, less reliance on photos, advancing concepts, longer sentences, and more complex special features.

★ **Blastoff! Universe**

Reading Level

Blastoff! Beginners — Grade K

Blastoff! Readers — Grades 1–3

Blastoff! Discovery — Grade 4

This edition first published in 2021 by Bellwether Media, Inc.

Library of Congress Cataloging-in-Publication Data

Names: Neuenfeldt, Elizabeth, author.
Title: Sea horse / by Elizabeth Neuenfeldt.
Description: Minneapolis, MN : Bellwether Media, 2021. | Series: Blastoff! reader : Animal life cycles | Includes bibliographical references and index. | Audience: Ages 5-8 | Audience: Grades K-1 |
Summary: "Relevant images match informative text in this introduction to the life cycle of a sea horse. Intended for students in kindergarten through third grade"-- Provided by publisher.
Identifiers: LCCN 2020036817 (print) | LCCN 2020036818 (ebook) | ISBN 9781644874127 (library binding) | ISBN 9781648340895 (ebook)
Subjects: LCSH: Sea horses--Life cycles--Juvenile literature.
Classification: LCC QL638.S9 N48 2021 (print) | LCC QL638.S9 (ebook) | DDC 597/.6798156--dc23
LC record available at https://lccn.loc.gov/2020036817
LC ebook record available at https://lccn.loc.gov/2020036818

Editor: Betsy Rathburn Designer: Jeffrey Kollock

Printed in the United States of America, North Mankato, MN.

Table of Contents

What Are Sea Horses?

Sea horses are tiny fish. They live in warm, **shallow** ocean waters.

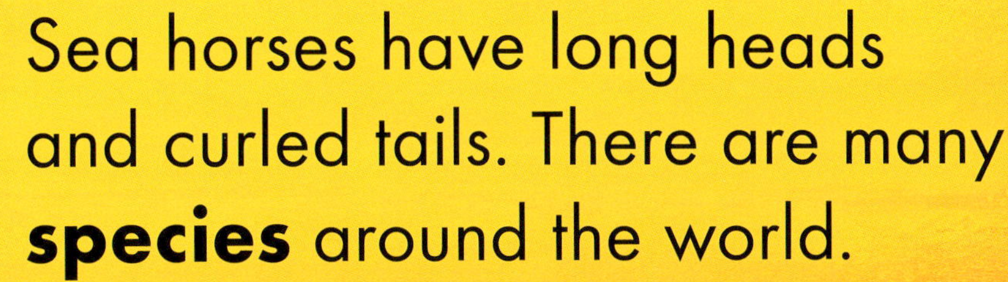

Sea horses have long heads and curled tails. There are many **species** around the world.

Life Cycle of a Sea Horse

The sea horse life cycle begins with **courtship**. Males and females greet each other.

They change colors and **synchronize** movements. This strengthens their bond.

courtship

Next, the two sea horses wrap their tails together.

They spin in circles. It looks like they are dancing!

Now it is time to **mate**.
Female sea horses give
their eggs to males.

Males store eggs in their **brood pouches**. Brood pouches can hold thousands of eggs!

The males are now **pregnant**. They carry the eggs for several weeks.

Then, eggs begin to **hatch**.
The brood pouch gets bigger!

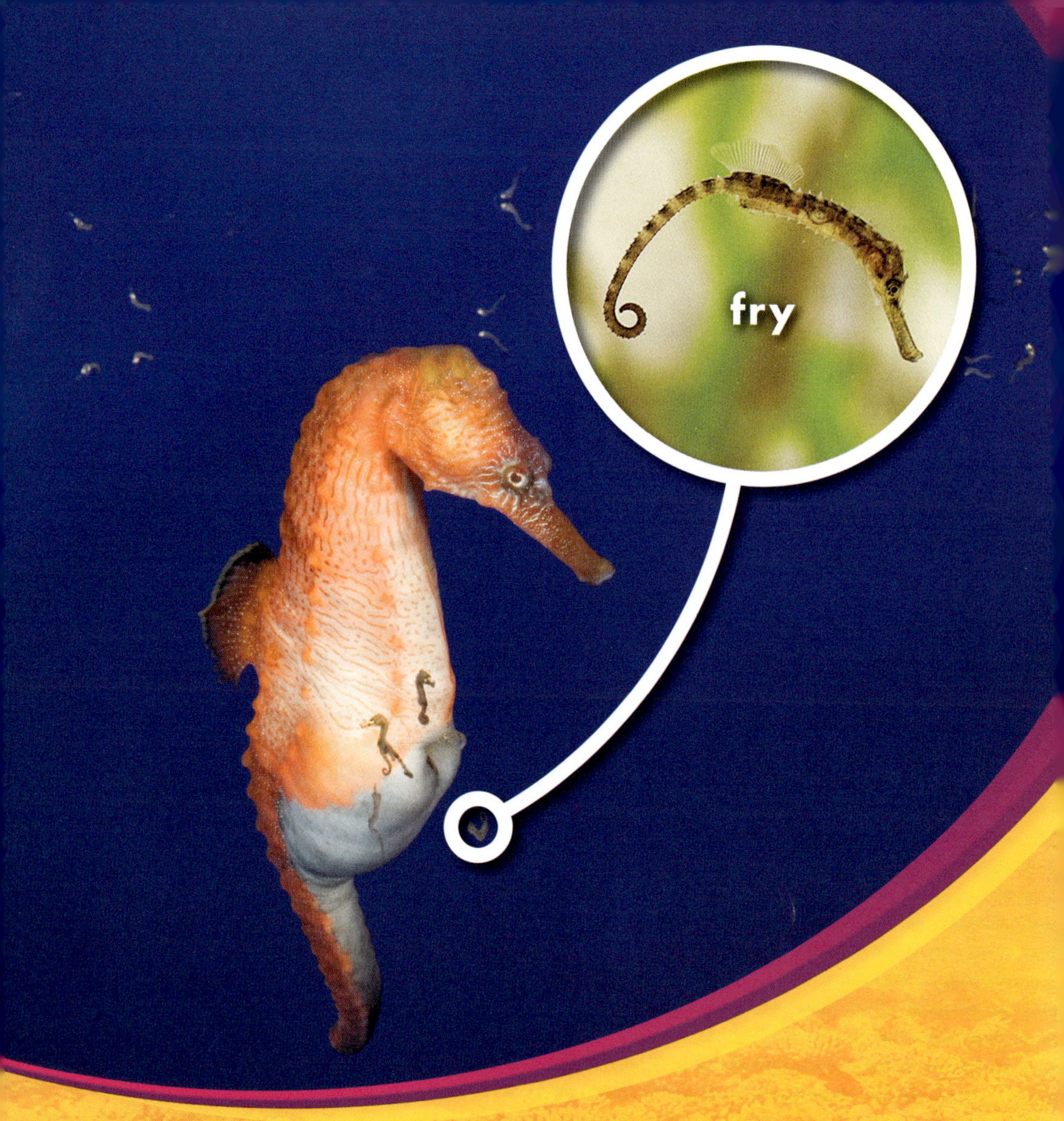

In the next stage, male sea horses give birth. Newborn **fry** leave the brood pouch.

Fry are small. They are less than 1 inch (2.5 centimeters) tall!

Sea horse parents
leave after fry are born.
Young fry are on their own.

Fry search for their own food. They eat newborn shrimp and **rotifers**.

All Grown Up!

In about 14 weeks, fry become adult sea horses. Some sea horses are tiny. Others grow over 1 foot (0.3 meters) long!

They eat shrimp and newborn fish.

In a few months, young sea horses are ready to mate. Once sea horses mate, they make fry together for life!

Life Cycle of a Sea Horse
Egg
1.
2.
Fry
3.
Adult

Glossary

brood pouches—pockets on male sea horse bodies that hold eggs

courtship—the act of seeking the love or companionship of someone

fry—young sea horses

hatch—to break out of an egg

mate—to join together to make young

pregnant—carrying one or more unborn babies

rotifers—very small animals that live in the water

shallow—not deep

species—types of an animal

synchronize—to cause things to happen at the same time and speed

To Learn More

AT THE LIBRARY

Bodden, Valerie. *Seahorses.* Mankato, Minn.: Creative Education, 2019.

Bozzo, Linda. *How Seahorses Grow Up.* New York, N.Y.: Enslow Publishing, 2020.

Leaf, Christina. *Sea Horses.* Minneapolis, Minn.: Bellwether Media, 2017.

ON THE WEB

FACTSURFER

Factsurfer.com gives you a safe, fun way to find more information.

1. Go to www.factsurfer.com.
2. Enter "sea horse" into the search box and click 🔍.
3. Select your book cover to see a list of related content.

Index

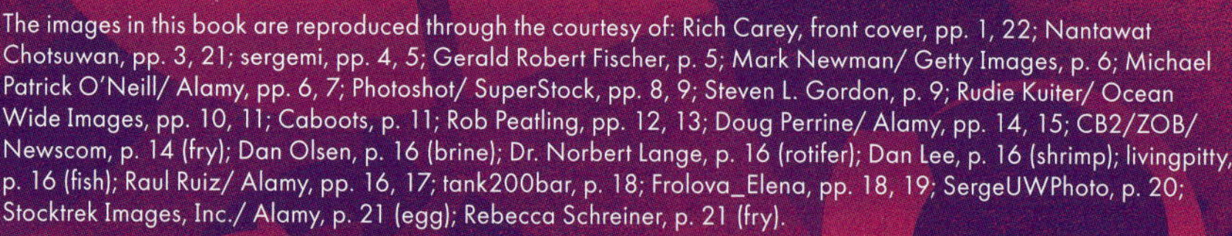

The images in this book are reproduced through the courtesy of: Rich Carey, front cover, pp. 1, 22; Nantawat Chotsuwan, pp. 3, 21; sergemi, pp. 4, 5; Gerald Robert Fischer, p. 5; Mark Newman/ Getty Images, p. 6; Michael Patrick O'Neill/ Alamy, pp. 6, 7; Photoshot/ SuperStock, pp. 8, 9; Steven L. Gordon, p. 9; Rudie Kuiter/ Ocean Wide Images, pp. 10, 11; Caboots, p. 11; Rob Peatling, pp. 12, 13; Doug Perrine/ Alamy, pp. 14, 15; CB2/ZOB/ Newscom, p. 14 (fry); Dan Olsen, p. 16 (brine); Dr. Norbert Lange, p. 16 (rotifer); Dan Lee, p. 16 (shrimp); livingpitty, p. 16 (fish); Raul Ruiz/ Alamy, pp. 16, 17; tank200bar, p. 18; Frolova_Elena, pp. 18, 19; SergeUWPhoto, p. 20; Stocktrek Images, Inc./ Alamy, p. 21 (egg); Rebecca Schreiner, p. 21 (fry).